T0363135

becoming aware

Daniel J. Siegel, MD, received his medical degree from Harvard University and completed his postgraduate medical education at UCLA, where he is currently a clinical professor. He is the executive director of the Mindsight Institute, and the author of numerous books, including the bestsellers *Mindsight* and *Brainstorm*, as well as *No-Drama Discipline* and *The Whole-Brain Child* (co-authored with Tina Payne Bryson). He lives in Los Angeles with his wife and occasionally with his launched adolescents.

becoming
aware

a 21-day
mindfulness program
for reducing anxiety
and cultivating calm

Daniel J. Siegel

SCRIBE
Melbourne • London

Scribe Publications
18–20 Edward St, Brunswick, Victoria 3056, Australia
2 John St, Clerkenwell, London, WC1N 2ES, United Kingdom

Published by Scribe 2022

This edition published by arrangement with TarcherPerigee, an
imprint of Penguin Publishing Group, a division of Penguin
Random House LLC

Copyright © Mind Your Brain, Inc. 2021

All rights reserved. Without limiting the rights under copyright
reserved above, no part of this publication may be reproduced,
stored in or introduced into a retrieval system, or transmitted,
in any form or by any means (electronic, mechanical, photocopying,
recording or otherwise) without the prior written permission
of the publishers of this book.

The moral rights of the author have been asserted.

Illustrations by Madeleine W. Siegel
Book design by Lorie Pagnozzi

Printed and bound in Australia by Griffin Press, part of Ovato

Scribe is committed to the sustainable use of natural resources and
the use of paper products made responsibly from those resources.

Scribe acknowledges Australia's First Nations peoples as the
traditional owners and custodians of this country, and we pay our
respects to their elders, past and present.

978 1 922585 12 7 (Australian edition)
978 1 914484 05 6 (UK edition)
978 1 922586 17 9 (ebook)

Catalogue records for this book are available from the
National Library of Australia and the British Library.

scribepublications.com.au
scribepublications.co.uk

contents

introduction

In this hands-on guide to the mindfulness practice I developed called "the Wheel of Awareness," I invite you to become a student of the architecture of your own mind. Mindful awareness, as you will see, involves more than just being simply aware—it is being aware of the aspects of the mind itself. Mindfulness in its most general sense is about waking up from a life in which we are set on "automatic," to becoming more aware of and sensitive to the novelty of our everyday experiences. Moving from automatic enables us to be more present in our lives and for the people we love.

In the pages of this book we will dive deep into three learnable skills that have been shown in carefully conducted scientific studies to support the cultivation of well-being:

- FOCUSED ATTENTION
- OPEN AWARENESS
- KIND INTENTION

When we develop focused attention, open awareness, and kind intention, research reveals that we will be better equipped to weather life's storms with a stronger sense of overall calm and happiness. Research also reveals that we can impact the health of our bodies in very real and measurable ways including:

1. Improving immune function to help fight infection.

2. Optimizing the level of the enzyme telomerase, which repairs and maintains the ends of your chromosomes, keeping your cells—and therefore you—youthful, functioning well, and healthy.

3. Enhancing the "epigenetic" regulation of genes to help prevent life-threatening inflammation.

4. Reduce stress and improve cardiovascular functions.

5. Help the brain grow more linkages among separate regions—something called "integration," which is the neural basis for resilience and well-being!

As you move through the 21-day program in this book, you will develop your own personalized Wheel of Awareness practice. It is my hope that, as your practice deepens, you will discover, as I and many others who work with the Wheel have, that you will experience more clarity and peace, a deeper sense of stability and even vitality, not just during the practice itself but during the rest of your day as well.

A Practical Tool

The Wheel is a very useful visual metaphor for the way the mind works. The concept came to me one day as I stood looking down at a circular table in my office. The tabletop consists of a clear glass center surrounded by a wooden outer rim. It occurred to me that our awareness could be seen as lying at the center of a circle—a hub, if you will—from which, at any given moment, we can choose to focus on a wide array of thoughts, images, feelings, and sensations circling us on the rim. In other words, what we could be aware of could be represented on the wooden rim; the experience of being aware could be placed in the hub.

The central hub of that table, of what we were now calling the Wheel of Awareness, represents the experience of being aware, of knowing that one is surveying the knowns of life. The rim came to represent that which is known; for instance, at this moment, you are aware of the words you are reading on this page, and now perhaps you've become aware of the associations you are having with the words—the images or memories that come to mind.

Consciousness involves our subjective sense of knowing—like your awareness now of my writing the word "hello." In this book, we will be guided by the perspective that consciousness includes both the knowing and the known. You know I wrote "hello." You

Basic Wheel

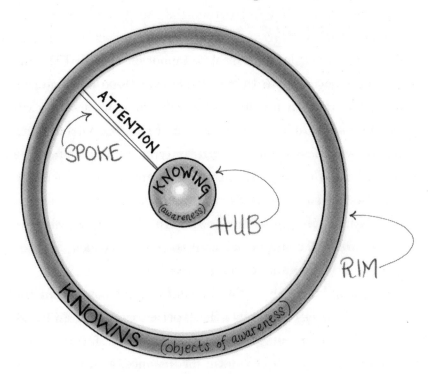

knowing is awareness; "hello" is the known. The knowing is in the hub; the knowns are on the rim.

Now imagine what might happen if, from the starting point of the hub, our attention was directed out to any of the various knowns on the rim, focused on one point or another—on a given thought, a perception, or a feeling, any single one of the wide range of knowns of life that rest on the rim of the Wheel. Extending the metaphor of the wheel, one might envision these moments of focusing attention as a spoke on the wheel. The spoke of attention connects the hub of knowing to the rim of the knowns.

In the practice, I invite my patients or students to imagine their minds to be like the Wheel. We envision next how the rim could be divided into four parts or segments, each of which contains a certain category of knowns. The first segment contains the category of knowns of our first five senses: hearing, sight, smell, taste, and touch; the second segment represents another category of knowns, one that includes the interior signals of the body, such as sensations from our muscles or from our lungs. The third segment contains the mental activities of feelings, thoughts, and memories, while the fourth holds our sense of connection to other people and to nature, our relational sense.

The practice of the Wheel of Awareness involves slowly moving the spoke of our attention around the rim, bringing into focus, one by one, each of the elements of that segment, and then moving the spoke of attention to the next segment. Systematically we take in rim element by rim element, moving the spoke of attention around the rim of knowns. By practicing the Wheel, we are able to widen and expand our ability to take in the vast sea of our own awareness at the same time as we are able to pinpoint certain elements and focus in on them. People who participate in the practice have remarked that the process feels like strengthening their minds.

WHEEL OF AWARENESS

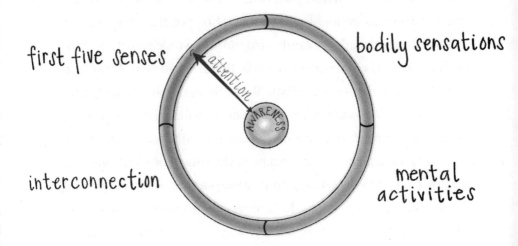

first five senses

bodily sensations

attention

AWARENESS

interconnection

mental
activities

Moving Toward Integration

The Wheel was designed as a practice that can balance our lives by integrating the experience of consciousness. How? By distinguishing the wide array of knowns on the rim from one another and from the knowing of awareness in the hub itself, we can differentiate the components of consciousness. Then, by systematically connecting these knowns of the rim to the knowing of the hub with the movement of the spoke of attention, it becomes possible to link the differentiated parts of consciousness. This is how, by differentiating and linking, the Wheel of Awareness practice integrates consciousness.

The essential idea behind the Wheel is to expand the container of consciousness and, in effect, balance the experience of consciousness itself. "Balance" is a common term that we can understand scientifically as coming from this process that we are calling "integration"—the allowing of things to be different or distinct from each other on the one hand, and then connecting them to each other on the other. When we differentiate and link, we integrate. We become balanced and coordinated in life when we create integration. Various scientific disciplines may use other terminology, but the concept is the same. Integration—the balancing of differentiation and linkage—is the basis for optimal regulation that enables us to flow between chaos and rigidity; integration is the core process that helps us flourish and thrive. Health comes from integration. On a more practical, day-to-day basis, what this means is moving from a less reactive place to a more receptive and calm outlook—moving from panic to presence. It's that simple, and that important. So now, without further ado, let us get to work!

Week One

laying the groundwork
with breath

I n Week One of this program, we will lay the groundwork for our Wheel practice by focusing on a most important building block: learning how to stabilize attention using our breath. There's a certain internal coherence that breath awareness creates, which is likely due to the repeating pattern of the inhalation and the exhalation, the in-breath and the out-breath. The anticipating of something followed by its arrival that breathing encompasses is deeply satisfying and grounding. It can give life a sense of being predictable and reliable. For many, focusing on the breath in this way creates coherence in the physiological balance of the heart as well as the clarity of the mind that can continue long after the practice period itself. Letting this practice of focusing on the breath and returning the focus to it when the mind becomes distracted can be one of your greatest tools in developing your practice, and a gift that keeps on giving as you go about your days because, indeed, we are always breathing.

In each moment of our waking lives, there is a constant flow of energy and information filtering through our minds. We can find a way to build our well-being so that we can be open to any objects of attention—sensation, memory, imagination—so that we do not have to exclude these from awareness. Imagine having a mind that can simply have a "bring it on" stance, open from the inside out to whatever life brings our way! One way to strengthen how we monitor this flow is to stabilize the lens with which we experience it. A very useful focus for this practice, one found in many cultures around the world, is the breath. When we do a basic breath-awareness practice, we are strengthening the monitoring capacity of the mind so that we stabilize attention. With the extension of the more elaborate Wheel practice itself, as we'll soon see, we will be furthering that stabilizing of attention and then also adding other aspects of strengthening to both the monitoring and the modifying of that energy flow.

In this first week of our journey, I invite you to alternate the two breath practices described below and each day record your impressions on the pages provided. Again, feel free to modify this and any part of the practice as we move ahead to accommodate your particular needs.

Breath Awareness Meditation

Some starting tips:

TRY TO STAY AWAKE. When you reflect inwardly, such as focusing on the breath as a sensation of the body, you are letting go of attention directed toward the outside world. For some, this inward focus is so different from an outward focus that it can feel unfamiliar, awkward, or even uncomfortable. Some people find this inner focus dull and boring. The tendency in this situation can be to lose focus, become less alert, get sleepy, and even to fall asleep. While napping is perhaps one of the most underrated of human activities, staying awake for the practice may be something you want to do to gain its benefits. Staying alert is in fact part of learning to strengthen the mind's focus of attention by noticing when you are getting groggy and then waking yourself up. Monitoring your state of alertness is part of learning to strengthen how you monitor energy and information flow. Now you can take this information about your sleepiness and modulate energy in ways to keep yourself awake, and even to become more alert.

For example, if you've closed your eyes, you might consider opening them a bit to let light in and stimulate your brain. You can do this whole practice with eyes wide open, or partially open with eyelids almost closed. If that doesn't work to keep you awake, try sitting up if you're lying down. If you're sitting down, you can try standing, and if you're standing, try walking around. You can do something to change energy flow and enliven the mind to stabilize attention. The key is to monitor your own state of energy and alertness and then do something about it. If you need a nap, sometimes it's just better to intentionally let the reflective practice go, for now, and simply let the nap happen and enjoy the rest!

A second tip: If you are doing this as a group practice, it may be helpful to establish some collective agreement that if someone does indeed go into the sleep state, and that person starts snoring, other members of the group have permission to wake up the snoring individual. It's really hard for others to ignore a snore. Better to agree ahead of time and give permission for a respectful, gentle prod to awaken the sleeper.

RELAXATION VERSUS REFLECTION. There is a difference between relaxation and reflection. Relaxation techniques are great for getting calm, but their outcomes have been shown to be quite different from the effects of a meditative mindfulness practice. So while you may get relaxed doing this reflective breath exercise, or later doing the Wheel practice, it is equally possible that you will not come to feel relaxed at all, and that's perfectly fine. Reflection is not the same as relaxation—neither in the doing, nor in the results. Reflection is more like becoming stable and clear, even in the face of a lot of chaos around you—or inside you. The state of mindful awareness is about monitoring with stability whatever is arising as it arises. That's the receptive awareness that we are calling "presence." This is the clarity that reflection builds as it enables things to arise and simply be experienced within awareness, the hub of our Wheel.

SENSING VERSUS OBSERVING. There is a difference between sensing something and observing it. When we open awareness to sensation, such as that of the breath, we become a conduit directing the flow of something into our awareness; for instance, enabling the sensation of the breath at the nostrils to flow into consciousness. *Attention* here is more like a hose letting the water flow through it. By contrast, when we *observe* something, there is a quality of being more like a witness constructing a perception—for example, freezing the water and then constructing an igloo out of the blocks of ice—rather than a conduit directing a stream. And as we'll see, when we begin to witness and narrate from that observational stance, we construct a story about something—even about the breath—rather than simply sensing the conduit flow of that sensory stream. If energy flow is like soapy water, then the mind is like the loop that can simply let bubbles emerge or shape them into symbols.

Observation is a gateway to being a witness and then becoming the narrator of an experience. If you like acronyms, as I do, this is how you OWN an experience: observe, witness, and narrate. These are all forms of construction in that there is a role we play as an observer, a witness, and a narrator, each contributing to the construction of an experience in that moment. This construction can be quite distinct from the sensing flow of being a conduit of experience, of what we can call "conduition."

The key to starting this reflective breath-awareness practice is to let sensation of the breath be the focus of attention and let it fill awareness. That is quite different from being invited to observe the breath, or witness it, or narrate the experience of breathing: "I am now breathing." This may perhaps sound like a subtle difference, but as you may come to see, distinguishing the difference between sensing and observing is a fundamental part of integrating your experience and empowering your mind.

BE KIND TO YOURSELF. These may be *simple* practices, but that does not make them *easy* ones. In many ways, reflecting inwardly is one of the biggest challenges we face as human beings. As the French mathematician Blaise Pascal said, "All of humanity's problems stem from man's inability to sit quietly in a room alone." Indeed, our ability to reflect lies at the very heart of emotional and social intelligence, skills many people have not learned. These are tools that will empower you to know your inner mind and connect with the inner, mental life of others.

We are so accustomed to focusing outwardly that such reflective practice is often quite new for many people. To sit quietly for any length of time feels unbearable for some. We love to be distracted by external stimuli or to speak and fill the gaps of silence in our lives. And so it is quite important to be gentle with yourself and realize that much of your life may have been focused on the external world and filled with input from your surroundings— from people, gadgets, and other things out in the environment around you. Now you are enriching your life's journey by learning to reflect on your inner life.

It can be frustrating at first to get comfortable with these reflective practices. Again, I invite you to be kind to yourself. This is hard work, and there is no way to do this "perfectly." Remember that your mind has a mind of its own. Part of your task is to realize that energy and information simply flow. Sometimes you can direct them well, guiding attention; sometimes they just take on a life of their own, as attention is pulled this way and that. Being open to whatever happens is the first step. Being kind to yourself as you travel through this guide will assist in that.

....

aware on-the-go:

Focus on Your Feet

In need of a quick centering tool you can utilize

anytime and anywhere? Close your eyes, take a

deep breath and simply . . . focus on your feet.

By placing all of your attention on your feet and

focusing in on how they root you to the earth,

you can quite literally "ground" yourself in a

matter of moments.

....

In this first week of our three-week program I'd like you to practice the following two exercises on alternating days. At the end of this section are spaces for you to record your experiences, make notes, and chart your progress as you sharpen your attention and expand your awareness using your breath.

Before we begin, it is important to note that, for some, the breath—or another particular focus of attention—may actually not be soothing either at first, or even after practice. For various reasons, including difficult experiences in the past that remain unresolved such as trauma, any meditative focus can reveal sensations that may create distress. If this becomes your experience, finding another focus during that part of the practice—such as on the sensations from the ends of the fingers or a light streaming through a window—can also serve as a useful focus of attention for this practice. The key is to be kind and caring to yourself, realizing we are each unique individuals, and adjust these suggested instructions throughout this or any reflective practice in order to help support your growth and well-being.

Exercise 1
Breath Awareness to Stabilize Attention
Find a quiet space free of interruptions. Take a moment to find a comfortable position—you can be seated, lying down, or standing. Turn off any gadgets that might disturb your five-minute practice. If you have a timer, set it to sound a gentle alarm at five minutes. If you are sitting on a chair, uncross your legs, have

your back straight but comfortable, and have both feet flat on the floor. If you are sitting on the floor, legs folded underneath you, let your back be straight and your body in a comfortable position that you can maintain for a few minutes. If you sometimes experience back pain, as I do, you might do this practice lying down, but be aware that you may be more likely to take a nap. One technique that helps me avoid falling asleep when lying down is to have one forearm lifted, elbow on the floor, with my hand raised up toward the ceiling. If you do fall asleep, you'll know because your arm will likely have fallen down onto your chest (and perhaps even wake you up).

Your eyes can remain open if you like, or you can let them partially close with a soft focus. Some may find it easier to simply let their eyes totally close to remove the sensory distraction of light.

Before closing your eyes, try these four steps:

1. Let your visual attention go to the middle of the room.

2. Now send your attention to the far wall (or ceiling, if you are lying down).

3. Next, bring your attention back to the middle of the room.

4. Finally, let your visual attention come to about the distance at which you might hold a book you are reading.

Take a moment to notice how you can determine where attention goes. Here, with your visual attention, you are simply directing the energy of light into awareness.

I invite you to read the following instructions, and as you do, you may use them as a guide to try the practice as you go. Then once you are familiar with the practice, you can go to the Resources tab on my website (DrDanSiegel.com/Resources) and listen to my voice as I guide you through this and later practices as well. Once you get familiar with a practice after first reading the full instructions, you can be guided by my voice from the website or try it on your own from memory.

Here are the instructions:

- Let attention focus on the breath, beginning with the sensation of air moving in and out of your nostrils. Let the sensations of the in-breath and the out-breath fill awareness. Just ride the wave of the breath, in and out.

- Focus on your chest, letting the sensation of the rising and falling fill awareness. In and out, in and out, ride the wave of the breath.

- Now let your attention move to your abdomen. If you've never done "belly breathing," you can put a hand on your abdomen and let the sensation of its movement fill awareness. As air fills the lungs, the diaphragm beneath them pulls down and pushes the abdomen outward; as the air escapes the lungs, the diaphragm relaxes and the abdomen moves inward. Continue to let the sensation of the abdomen moving in and out fill awareness. Ride the wave of the breath in and out, letting awareness be filled with the sensation of the abdomen's movement.

- Now let attention find the sensation of the breath wherever it feels most natural for you. It may be the sensation of the abdomen moving in and out; it may be the chest rising and falling; it may be the sensation of air moving in and out of the nostrils. Or perhaps it's the whole body just breathing, in and out. Wherever the sensation of the breath is felt most readily for you, let that become the focus of attention.

- Now let the sensation of the breath fill awareness. In and out, in and out, ride the wave of the breath, in and out. At some point, awareness may become filled with something other than the breath. When you realize that awareness is no longer with your breath, redirect your attention to the sensation of the breath.

- Continue to focus on the breath for a few cycles, refocusing on it whenever a distraction has taken your focus away from the breath, and see how this goes. If you are reading these instructions as you practice, you may like to close your eyes for a few cycles of breath before continuing to read.

Where attention goes, neural firing flows, and neural connection grows.

aware on-the-go:

Box-Breathing

We are always breathing and yet—in our often fast-paced and cyber-speed lives—it can be easy to forget this fundamental biological process happening inside of our bodies. Here's a simple and quick way to connect to your breath and expand your awareness anytime and anywhere.

1. Sit or stand upright and slowly exhale through your mouth, releasing all of the oxygen in your lungs.
2. Slowly inhale through your nose to the count of four.
3. Hold your breath for another slow count of four.
4. Exhale through your mouth for another slow and steady count of four.

Exercise 2

Now let's try adding one more component. For some people, finding a general word that represents the distraction that pulls attention away from the breath can be helpful. If a thought took
your attention away from streaming the sensation of the breath
into awareness, especially if it was a thought that came back re-
peatedly, you might like to try saying quietly in your inner
mind, "Thinking, thinking, thinking." For some, this naming
of a distraction helps to let it go and eases the ability to redirect
attention to the sensation of the breath. Similarly, if a memory
takes over awareness and replaces the breath, then saying inter-
nally, "Remembering, remembering, remembering," can be helpful to redirect attention away from the memory and return it to
the sensation of the breath. For others, this naming process is
itself too distracting and not really helpful. For them, it is more
straightforward to simply take note of the distraction without
naming it and then redirect attention to the sensation of the
breath.

In addition to labeling or noting distractions—then returning
to the breath—remember to try to bring kindness to this experience. It may be helpful to consider this perspective: The breath
practice is like contracting and relaxing a muscle during exercise.
Focusing on the breath is contracting the muscle; the inevitable
distraction is relaxing the muscle. You don't need to create the
distractions—they will happen naturally, as the mind has a mind

of its own! But you can intentionally create an attitude of kindness when these distractions come, being open to whatever arises, observing the distraction, realizing it is an object or activity of the mind, and then returning the focus of attention back to the breath—allowing your kindness to frame this process with a gentle, nonjudgmental attitude. This is how you can use kindness with the mindsight tripod of openness, observation, and objectivity.

If you were to only be in the conduit flow of whatever was happening in sensation, then getting lost in a distraction would just be your flowing sensory experience. In this case, you'd only be harnessing the "O" of openness of your mindsight lens. Instead, stabilizing attention enables us to be in the flow of the sensation of the breath—open to the conduition flow—and then use the construction tools of the mind's capacity for observation and objectivity so we note the new thought or memory as a distraction and do not just flow with it; then we can construct the redirecting process to get attention back to the sensation of the breath. In broad terms, this simple breath-awareness practice invites us to be open to the flow of the breath, observe when that focus of attention has wandered, and objectively move attention back to the breath. That's the integration of differentiating openness, observation, and objectivity, and then linking them together as we stabilize attention.

So let's try this basic breath-awareness practice again, this time with the invitation to either label or simply take note of distractions and kindly return, again and again, to the breath. Remember, if you'd like to hear my voice guide you through this breath practice, please go to my website (DrDanSiegel.com /Resources/Everyday-Mindsight-Tools).

If you're doing this on your own steam and this is your first time doing this mindfulness of the breath practice, set a timer for three minutes. You may like to consider what type of sound the timer is set to make—one that might be different from what you use to awaken from sleep in the morning. If you've done this before, give five minutes or more a try. Once your timer is set, let yourself sense the breath, refocusing when a distraction has filled awareness with something other than the breath, and then continue to ride the wave of the breath, in and out, until the timer lets you know it is time to stop. Before starting the timer for any reflective practice, find a comfortable position, back straight, in a space in which you will not be interrupted.

Ready? Enjoy the ride!

After the timer signals it is time to stop, you may feel calm or energized, refreshed or tired. If you are in a challenging period in your life, you may even feel more anxious or tense, as spending time dwelling on our interior can also make us more aware of the difficulties we are facing. Recall that this is an exercise. Doing an exercise does not mean we have to feel a certain way afterward, or even that we will feel the same way each time we try it. Why is this considered an exercise? It is an exercise because you are strengthening your capacity to focus attention, to notice a distraction that is not salient or relevant to the task at hand—a noticing scientists call "salience monitoring"—and then to redirect attention intentionally. There are different brain circuits for each of these facets of attention—sustaining focus, noticing, and redirecting—and you are training each of them.

Keep in mind our basic statement: Where attention goes, neural firing flows, and neural connection grows. You've been activating several important parts of your brain in just a few short minutes of practice!

In other reflective exercises that will be part of our Wheel practice, we will explore and expand the capacity for open awareness, or open monitoring—meaning, letting things simply arise, and being in an open, receptive state. This open awareness, along with the fundamental elements of attention practice—sustaining, noticing, and redirecting—will each grow stronger as your practice deepens.

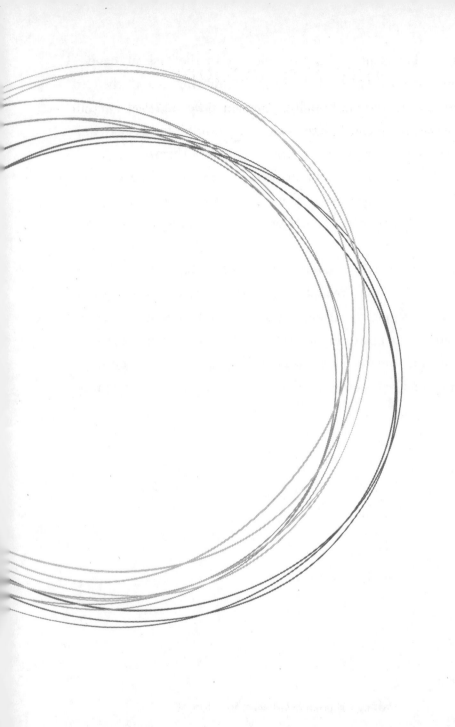

day 1 reflections

How was that for you? Take a moment now to reflect on your experience with your breath so far.

day 2 reflections

Following your breath meditation, how did you interact with your breath throughout the rest of the day/evening? Were you more conscious of inhaling and exhaling? Did it impact your emotional state? How?

day 3 reflections

When you are doing your breath practice, does any particular imagery enter your mind? Describe a few of the images. What have been the most challenging aspects of the practice so far? Do you find being kind and patient with yourself difficult, or do these compassionate ways of being inwardly come easily?

day 4 reflections

Now having practiced these breath awareness exercises for several days, how are you feeling the remainder of the day? More centered? Calm? Describe.

day 5 reflections

As we move closer to the Wheel practice in Week Two, I now invite you to reflect a bit on the four rim segments and note now how often you found yourself on each segment during today's breath-awareness practice—were you on one more than others? What specific thoughts or feelings did you experience? This is how we may be SIFTing our minds as we experience sensations, images, feelings, and thoughts in our lives.

The Five Senses: Hearing, seeing, smelling, tasting, touching

Bodily Sensations: Signals from muscles and bones, and from our bodily organs like the heart or intestines

Mental Activities: Emotions, memories, thoughts, beliefs, moods, attitudes, hopes, longings, desires, and more

Interconnection: A sense of connection to other people and to the planet—to nature and the environment surrounding our body

day 6 reflections

Were there any moments in your practice today in which you felt all or partially removed from the practice? Was it an immediate break, or did you slowly become aware that you were pondering various distractions, thoughts, or worries? Describe these moments and how you were able to bring yourself back into the practice. Are there any tools you can use as you move deeper inside with the Wheel practice in the coming pages?

day 7 reflections

Three Pillars of Mind Training

Reviews of research on mind training suggest that three factors help us to create well-being and happiness in our lives—focused attention, open awareness, and compassion, or what we might call "kind intention." *Focused attention* involves our capacity to sustain our concentration, refocusing our attention on our intended object of attention. *Open awareness* is the ability to remain receptive to objects within awareness but not get attached to them or lost in them. *Kind intention* involves a state of mind in which we feel positive regard, compassion, and love for ourselves and others around us. For today's reflection, record any moments during this week's practice in which you felt each one of these pillars come into play in your practice. What thoughts, memories, or images emerged, either during the practice or outside of it?

Focused attention:

Open awareness:

Kind intention:

Week Two

exploring the wheel

I n this second week of our journey through the architecture of our own minds, we will dive right into the Wheel and explore how it can serve as both a metaphor and a model for our innate capacity to focus attention and expand our awareness. This week I invite you to try out the Basic Wheel practice each day of the week. In Week Three of the program, we will draw upon this basic orientation to the Wheel practice and expand upon it with the Full Wheel.

The Basic Wheel of Awareness

Find a quiet place where you'll be able to sit, lie down, or stand for about half an hour, uninterrupted. Turn off your gadgets. Take another look at the diagram of the Wheel on the next page and

Basic Wheel

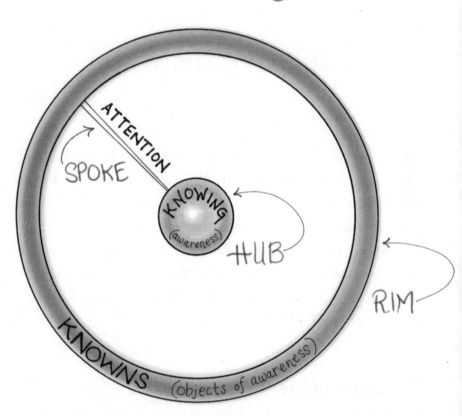

form a mental picture of it in your mind, if possible; or simply knowing its components is fine, too. Recall that the hub represents the knowing of awareness; the rim, the knowns; and the spoke, the focus of attention. (Again, you may find it useful to read through the entire exercise before trying out the practice, and then guide yourself from memory. Alternatively, you can go to the Resources section on my website, DrDanSiegel.com/Resources, if you are interested in being guided by my voice through the practice.)

Let's begin by focusing on the breath as a way to become centered. Just let the sensation of the breath fill awareness. Now shift your attention away from your breath. Imagine yourself in the center of the Wheel, in the hub of knowing, of being aware. Imagine sending a spoke of attention out from the hub of knowing to the first segment of the rim. Let's begin with focusing attention on the sensation of hearing, letting your awareness become filled with sound. . . .

(Staying with each sensation for about fifteen to thirty seconds can be a helpful duration.)

Now letting hearing go, imagine moving the spoke of attention over a bit on this first segment to the sense of sight, letting awareness become filled with light (through closed eyelids, or opening your eyes for a soft focus letting light in). . . .

Now move the spoke of attention over, letting light go and moving to the sense of smell, letting aromas fill awareness. . . .

Now move the spoke of attention over, opening to the sense of taste, letting tastes fill awareness. . . .

Now let the spoke of attention move over to the sense of touch, letting awareness become filled with the sensation of skin touching skin (hand in hand), skin touching clothing, skin touching the floor. . . .

Taking a deeper breath, imagine moving the spoke of attention over to the next segment of the rim, which represents the interior of the body—the sensations of the muscles and bones, and of the internal organs. (Here the timing for each part of the body mentioned will vary, anywhere from a few seconds to fifteen seconds or so.) Let's begin with the facial region, letting the sensations of the muscles and bones of the face fill awareness, moving attention up to the forehead and top of the scalp, now down the sides of the scalp passing by the ears, and now to the muscles and bones of the throat and neck. Now moving attention to the shoulders, and then streaming attention down both arms to the ends of the fingers

now bringing attention to the upper back and chest

now the lower back and the muscles of the abdomen

and now focusing attention on the sensations of the hips

and now streaming attention down both legs,

to the ends of the toes.

And now focusing attention on the pelvic region. Opening awareness to the sensations of the genitals . . . and now focusing attention on the sensations of the intestines, beginning deep in the abdomen with the lower intestines . . . and now moving upward toward the stomach region at the top of the abdomen . . . and now following those gut sensations up from the stomach through the center of the chest, opening to the sensations of the esophagus connecting the stomach to the throat and the interior of the mouth.

Now moving to the respiratory system, beginning behind the cheekbones with the sensations of the sinuses . . . then to the nose . . . and to the mouth . . . and then down the front of the throat to the trachea, the tube that brings life-giving air down into the center of the lungs, in the interior of the chest . . . the lungs on both sides expanding and contracting . . .

Now let the focus of attention move to the heart region, opening awareness to the sensations of the heart.

Now let the sensations of the whole of the interior of the body fill awareness, from head to toe. Knowing that science has now shown what wisdom traditions have known for a long time, that opening awareness to the sensations of the body is a powerful source of wisdom and intuition, I invite you to take a deeper breath, knowing you can always return to exploring this sixth sense of the sensations of the body, and now move the spoke of attention over to the next segment of the rim.

We will now direct the spoke of attention to focus on the third segment of the rim, the segment that represents the mental activities of emotions, thoughts, memories, beliefs, intentions, hopes, and dreams. I encourage you to invite any mental activities—thoughts, feelings, memories—to come into the hub of knowing. Just be open to whatever arises from the rim—or doesn't arise. There is no right or wrong. Many things may come, or nothing may come. Simply open the hub to whatever might arise from the rim of mental activities. Let's begin that practice right now (continue for about a minute and a half).

Next, again being open to anything from mental activities into the hub of knowing, I invite you to pay special attention to how a mental activity, such as a thought, first arises in awareness. Does it arise suddenly or gradually? Once it has presented itself to awareness, how does it stay present? Is it solid? Vibrating? And how does the mental activity, the thought, memory, or emotion, leave awareness? Does it leave from one "place" or another? Gradually or suddenly? Does it just get replaced by one mental activity or another, like a thought, a feeling, or a memory? And if it doesn't immediately get replaced by another mental activity, what does the gap feel like between one mental activity and the next? Here, I invite you to become a student of the architecture of mental life, studying how mental activities first present to awareness, stay present, and then leave awareness. Let's begin that practice, right now (continue for a minute and a half).

(Note: If you were doing the Full Wheel practice, this is when you would do the hub-in-hub bending or retracting of the spoke portion of the practice, which we will dive into in a later chapter.)

I invite you now to take a deeper breath. And now imagine moving the spoke of attention over to the fourth and final segment of the rim. This is the part of the rim that represents our relational sense, our connections to people and the world outside these bodies we were born into.

With the spoke of attention on this fourth segment of the rim, the relational segment, let awareness become filled with the sense of connection to people physically closest to you, right now. Now open to the sense of connection to family and friends not immediately next to you. Now let awareness become filled with a sense of connection to people with whom you work—in school, at work, in your community. Now open to the sense of connection to people who live in your neighborhood, to a sense of connection to people who share your

community . . . to people who live in your town or city, and now

opening to a sense of connection to people who live in your region or state . . . and to people who live in your country, to people with

whom you share your continent. And now see if you can open your sense of connection to all people who live on this precious planet, this place we've named Earth.

And now see if you can expand that sense of connection to all living beings on Earth. . . .

(In the Full Wheel practice, this is when you would add the statements of kind intention.)

I invite you now to find the breath again and ride the wave of the breath, in and out. Now, taking a more intentional and perhaps deeper breath, I invite you to let your eyes get ready to open if they've been closed, and we'll let this Wheel of Awareness practice come to a close for now.

A mother once said to me as I taught this practice to her and her adolescent son, "I never knew I was more than my thoughts or feelings." For her, distinguishing hub from rim was a revolutionary experience that empowered her to engage with life in a much more rich and nuanced way.

aware on-the-go:

Name It to Frame It

Ironically, some of our most deeply felt emotions can occur without our ever being quite aware of their origins or nature. They simply come and then they go. One tool for developing a greater capacity for awareness is to simply stop yourself in the moment and ask yourself, "What am I feeling?" In naming our emotions—sadness, confusion, fear, jealousy, love—we are able to be more present in our lives and form a deeper understanding of ourselves. We can "frame" the emotion to see it as if in a picture and take the time we need to sense it and understand it more fully. We become more aware.

day 8—reflecting on mind: your experience of the basic wheel

You've just completed the basic Wheel of Awareness practice. How was that for you? Did you find it challenging to hold your attention on the various segments of the rim? What was it like to return your attention—again and again—from the center of your wheel, letting your awareness again be filled with what the spoke was intending to bring in?

day 9—reflecting on the five senses

In the first segment, the five senses, what did sound feel like? Did you notice any shift in the quality of hearing when sound was selected as the singular focus in that moment? What did light feel like as you focused on visual sensations? How did color and contrast seem to you as vision was differentiated from other senses? With olfaction as the focus, what did aromas feel like in awareness? Were they more difficult or easier to sense than the other sensations so far? How did it feel to then move to taste? Did you notice your mouth or tongue moving to increase the sensations of taste? And what was it like to scan the body's skin in becoming aware of the sensation of touch? Did certain areas feel more sensitive than others?

For many people, taking this time to differentiate the first five senses from one another enables a heightened experience of being aware of each sensory stream. With practice, this capacity to sense with more clarity and detail will begin to enhance your enjoyment of daily life, bringing more intensity and pleasure, more vitality to your everyday experiences.

day 10—reflecting on the second segment of the wheel

Focusing on the second segment of the rim, what did turning your attention inward feel like? How did being invited to focus on the sensations of the muscles and bones feel? Did you become aware of sensations that perhaps were new to your awareness? Sensations exist even outside awareness, but they only become a part of our subjective experience when we bring them into consciousness with focal attention, guided or pulled. How did it feel to move attention throughout the body's muscles and bones? Your organs, intestines, and respiratory system? What was it like to guide attention to these sensations? Were any areas more challenging than others? Remember that this practice may be simple, but that does not make it easy! And sometimes certain bodily sensations may feel difficult to place attention on, or even uncomfortable in various ways. It's also helpful to remember to be kind to yourself, being patient and understanding that we each are different and can care for any experiences that are particularly challenging. Instead of seeing these as difficulties, it may be helpful to view them as opportunities to learn more about yourself, an invitation to grow in helpful ways. Taking time to write things down here, or in a journal; or to share with a trusted individual, a friend, therapist, or family member, can also be a useful step if issues that arise feel particularly confusing or distressing.

day 11—reflection: a focus on your heart

Focusing attention on the heart area—even if you aren't actually aware of the sensations of your heart beating—has been shown in some studies to help calm the mind, as it coordinates our brain's control of what is called the "autonomic nervous system." As we saw in Week One, this can be true, too, for becoming aware of the cycles of the breath. Research reveals that, with these aspects of bodily awareness, we are able to balance the equivalent of the brain's accelerator and brakes so that we learn to "drive the car" of the body in a smoother, more coordinated manner. Would you want to press on the brakes and accelerator at the same time? No—you'd want to coordinate these distinct slowing-down and speeding-up control functions. Focusing on the heart region or the breath gives us coordination and balance of the body and helps stabilize the mind. Following today's practice of the Wheel, did you note any moments where you were able to slow down and experience greater awareness of all the many layers of your experience in a given moment without becoming stressed or conflicted? *Interoception* is the sense we have either consciously or unconsciously of the internal state of our bodies. Studies suggest that people who have more interoceptive abilities have more capacity for insight and empathy, as well as emotional balance and intuition. And so building these body-awareness skills is a direct route to a deeper connection to our inner and interpersonal lives.

day 12—resting your attention on the mind

What was it like for you today to pay special attention to the dynamics of mental activities—how they first arose, stayed present, and then left consciousness? This can be especially challenging for many, as it seems to involve both an openness to whatever arises (as in the first part of this segment review), but then also a special focus on the nature of comings and goings. For some, the gap between mental activities is especially intriguing, as this space between, say, two thoughts, or memories, or emotions, has a very unusual quality that is new for many people to suddenly become aware of. For many of us who have not yet experienced a meditation or reflective practice such as the Wheel, it can be quite revelatory to be aware of the more subtle details of mental activity. Often, our previous experience may have been that one mental activity or another has dominated our awareness as chattering of the mind's activities, the ever-present stream of thoughts, and we haven't had the opportunity to experience that we in fact are more than our mind's chatter. With this mental activities third segment review, a new experience of open awareness—the hub—distinguishes itself even more from the elements of the rim. For many, this new awareness can be life changing and, literally, mind-altering—even mind-boggling. This is the beginning of more fully distinguishing the knowing from the knowns.

day 13—inhabiting our relational sense on the fourth segment of the rim

When you turned attention to the fourth segment of the rim today, what did that relational sense, our eighth sense, feel like for you? What was it like to let attention to mental activities go and shift to a focus on connections with others? This eighth relational sense can be experienced by some as confusing, as they are not sure what exactly they are focusing on. For others, a deep sense of love, of peace, of kindness and connection arises and they become filled with tears of joy and gratitude. Whatever your experience is, that is your experience. What was your experience like today?

day 14—mapping your wheel

In the space below, I invite you to list under each section of the Wheel, two lists; perhaps you can use different color pens or pencils for each. First, what came to mind as you moved through the sensory segment of the rim during today's practice? Next, record what, on a typical day in your life as you go about your various activities, your mind might rest upon in terms of the sensory input you receive—sounds, sights, smells, taste, touch. For example, under the five senses, the smell of your morning coffee as you take your first sip or the sounds of children playing in the street next to your home. In the following reflections section, it can be helpful to write down what the similarities and differences are of the different textures and feelings of your experience of the senses during your Wheel practice versus in your everyday life. Did the differences become more or less marked as you moved through Week Two and deepened your Wheel practice?

WHEEL OF AWARENESS

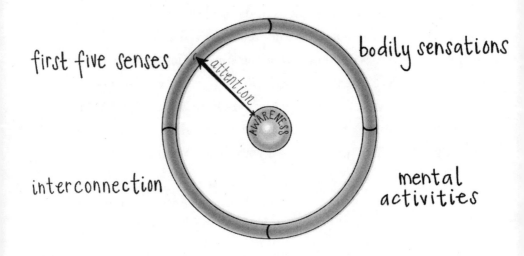

first five senses

attention

bodily sensations

AWARENESS

interconnection

mental activities

Reflections:

Week Three

the full wheel

Welcome to this final week of your introduction to the Wheel of Awareness practice. We will expand the practice with the Full Wheel. I encourage you to do the Full Wheel every day this week. Each day, read the outline below and commit it to memory before beginning your practice. Alternatively, you can go to the Resources section on my website, DrDanSiegel.com/Resources, if you are interested in being guided by my voice through the practice. This section contains two pages to reflect upon each day's practice, as well as a daily exercise that will help you further deepen your Wheel practice.

The Full Wheel Practice

The Full Wheel of Awareness practice can be outlined this way:

1. BREATH: Start with the breath to anchor attention and get grounded for the Wheel practice.

2. FIRST FIVE SENSES ON THE FIRST RIM SEGMENT: Let go of the breath as a focus of attention and begin the focus on the first segment of the rim—the first five senses, attending to one sense at a time: hearing, sight, smell, taste, touch.

3. INTEROCEPTION ON THE SECOND RIM SEGMENT: *Interoception* is the sense we have, either consciously or unconsciously, of the internal state of our bodies. Take a deep breath and move the spoke over to the second segment of the rim, which represents the internal signals of the body, and begin to "scan" your internal body. Systematically move the spoke of attention around the body, beginning with the sensations of the muscles and bones of the facial region, then moving on, one at a time, to the sensations of the head, neck, shoulders, arms, upper back and chest, lower back and muscles of the abdomen, hips, legs, and pelvic region. Now move to the sensations of the genitals, intestines, respiratory system, heart, and whole body.

4. MENTAL ACTIVITIES ON THE THIRD RIM
   SEGMENT: Take a deep breath and move the spoke
   over to the third segment of the rim, which
   represents mental activities. First part: Invite any
   mental activity—feeling, thought, memory,
   whatever—into awareness. Many things may arise,
   or nothing may arise; whatever happens is fine.
   Second part: Again, invite anything into awareness,
   but this time pay special attention to the way
   mental activities first arise, stay present, and then
   leave awareness. If a mental activity is not
   immediately replaced by another activity, what
   does the gap feel like before a new one arises?

5. HUB-IN-HUB WITH AWARENESS OF
   AWARENESS: Before we move the spoke of
   attention over to the fourth and final segment of
   the rim, we will explore the hub itself. In other
   words, we will strengthen our ability to be aware of
   awareness, to simply be aware yet not aware of
   something in particular. This can be accomplished
   by imagining the bending of the spoke of attention
   around, so it aims itself back into the hub; some
   prefer the image of retracting the spoke or simply
   leaving the spoke of attention in the hub of
   awareness—or having no spoke at all, just a wide-
   open hub. Whichever notion or visual image works

best for you, the idea of this part of the practice is the same: awareness of awareness itself, a dropping into the hub and simply being aware (let a minute or more pass on the first practice, extending perhaps to two or three minutes in later practices). Find the breath again and ride its wave, in and out. You can now get ready to straighten and extend the spoke to the fourth and final segment of the rim, our relational sense.

6. RELATIONAL SENSE OF OUR FOURTH RIM SEGMENT: In this final segment of the rim review, we'll explore our connections with other people and things outside of these bodies we were born into. Let's begin with a sense of connection to those physically closest to you right now. Next, open to the connection to friends and family; to a connection to people you work with; to people who live in your neighborhood; who share your community . . . who live in your city. . . .

Open to a sense of connection to those who share your state or region . . . to people who live in your country . . . Now open to a sense of connection to all people who live on Earth; and now see if you can open that sense of connection to all living beings on Earth. . . .

7. STATEMENTS OF KIND INTENTION: Now, knowing that science has recently revealed what wisdom traditions have known for many years—that cultivating intentions of kindness, care, empathy, and compassion can bring positive changes into our inner and interpersonal worlds—I invite you to repeat the following phrases silently, in your inner mind. We'll begin with short, basic statements of kindness, and then move on to these same intentions stated in a more elaborate way.

May all living beings . . . be happy. May all living beings . . . be as healthy as they can be. May all living beings . . . be safe.

May all living beings . . . flourish and thrive.

Now, taking a deeper breath, we send those same wishes, a bit more elaborated, to an inner sense of who we are, to a Me or I:

May I . . . be happy and live with meaning, connection, and equanimity, and a playful, grateful, and joyful heart.

May I . . . be as healthy as I can be and have a body that gives energy and flexibility, strength and stability.

May I . . . be safe and protected from all sorts of inner and outer harm.

May I . . . flourish and thrive and live with the ease of well-being.

Now, again taking a bit of a deeper breath, we will send those same elaborated wishes to an integrated sense of who we are. Combining our inner Me with our interconnected We, we continue with statements of kind intention for MWe:

May MWe . . . be happy and live with meaning, connection, and equanimity, and a playful, grateful, and joyful heart.

May MWe . . . be as healthy as MWe can be and have bodies that give energy and flexibility, strength and stability.

May MWe . . . be safe and protected from all sorts of inner and outer harm.

May MWe . . . flourish and thrive and live with the ease of well-being.

I invite you to once again find the breath and ride the wave of the breath, in and out.

And now, letting your eyes come open if they are closed, we'll bring this Wheel of Awareness practice to a close.

aware on-the-go:

Monkey Mind

Many meditation teachers and others have spoken of Monkey Mind—the busy, frenetic mind that is constantly chattering away, making it difficult to focus or stay centered. I find Monkey Mind a helpful metaphor for the mind's chatter in that it allows me to feel affectionate toward this tendency of my mind to dart around. Monkeys are quite adorable, after all. Next time you find yourself flooded with thoughts or emotions, take a moment to go easy on yourself and smile affectionately at your magnificent Monkey Mind.

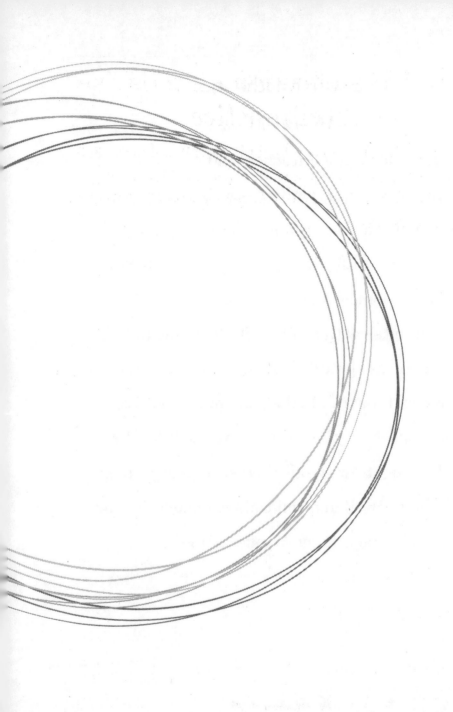

day 15—reflection on today's wheel practice

How did it feel today to move from the Basic Wheel to the Full Wheel practice? Did you notice that you were distracted less often? Did anything enter your mind uninvited? Images? Words? Feelings or emotions?

day 15—exercise & reflection: working with focal and non-focal attention

Exercise:

One important way to distinguish different forms of attention is to determine whether the stream of energy that is the focus of our attention enters awareness. If the focus of attention involves consciousness, it is called *focal attention*; if it does not, it is *non-focal attention*. In order to better understand this difference, take a moment to try the following short activity: Simply move around the room you are presently in. As you do so, notice what you are aware of as you sense and observe what is in front of your eyes, what you feel with your feet, or hands if you are moving in a wheelchair, or, if you are blind, what you feel with your cane or your hands as you move around. Take in as many of the signals from the outside world as you can, and bring them into the awareness of your consciousness. This is the knowing of being aware, and the knowns are what you are aware of. In other words, be as aware as you possibly can of your surroundings. Place the "spotlight of attention," like the beam of a flashlight along a dark pathway, on whatever you can as you move around the room.

The spotlight of focal attention aims your mental ability to focus energy flow into awareness. That is focal attention filling consciousness with certain aspects of your moving-around-the-room experience. At the very same time, studies reveal, your mind is also focusing a perhaps broader spotlight of attention on many aspects of your experience that never enter awareness. We call this *non-focal attention*. For example, on this journey, you were attending non-focally to your balance so you would not fall over, and you were attending to the space around you so you wouldn't bump into something as you moved around the room. You may have found that during this exercise you became absorbed in some thought or memory. At that moment, your focal attention was on these mental processes and no longer on your surroundings. But you didn't fall down or bump into something because your non-focal attention was taking care of attending to those potentially hazardous obstacles and keeping you safe—even without your awareness. Our nonconscious mind has a profound impact on what we do with our behavior, and on how we feel and think, even when we are not aware of these influences of non-focal attention on our mental lives.

Reflections:

Reflecting on this exercise, think about other scenarios in which you are both aware and not aware of your surroundings. For example, if you are walking along a hiking trail, you may pay attention to the rocks on the path ahead, disregarding the stones currently underfoot. Attention helps you survive; it helps you navigate the world in which you live. If you didn't pay attention, focal or non-focal, you could trip and fall. If you do pay attention, you're more likely to survive and thrive.

day 16—reflection on today's wheel practice

There are often subtle differences in our practice from day to day. Reflect on any differences between yesterday's practice and today's practice.

day 16—exercise & reflection: monitoring attention and awareness

Our days are filled with a combination of both guided and pulled attention. Sometimes we choose what to pay attention to, and sometimes the world's circumstances pull our attention, directing where the flashlight of attention is aimed. Interestingly, we need both guided and pulled attention, and we need both focal and non-focal attention. Imagine again our rocky hiking trail. We need to intentionally guide our attention, to the path itself so that we do not trip over a rock and fall. But if a bear suddenly crosses our path, we need to be able to have our attention pulled to this new fact of our experience (and quickly!). As we navigate this world, we must be nimble in terms of guided or pulled attention. And yet when it comes to the more day-to-day experience of living our lives—in other words, when there is no bear appearing on our path—our salience monitoring automatically evaluates what is significant enough for us to focus our attention on moment by moment, and this is usually happening without our even being conscious that these appraisals are being made by our nonconscious mind.

Here's an example to illustrate this important distinction. Imagine that you find yourself preoccupied with a challenging conversation you had with a friend last week. Without your even realizing it, feelings of sadness or anger that may have arisen at that time can be easily triggered now because salience has encouraged you to place significance on any scenario that feels intuitively related to the upsetting conversation you had with your friend. These emotions are now more relevant, more activated or "primed," because of the quarrel—even if they are not in your awareness at the moment.

This illustrative case of a disagreement helps illuminate the distinctions between awareness and attention. We engage in non-focal attention all the time. This is how our mind processes and keeps track of important things without using up the relatively limited mental space of awareness. This mental space of knowing, the subjective experience of being aware, can only work on a few items of information at a time—like a chalkboard of the mind, sometimes called *working memory*, that allows us to manipulate information and create new combinations consciously. Yet information processing does not require consciousness, and so we can imagine and calculate and come up with solutions to problems without using up this limited working memory space. To avoid flooding that space, our non-focal attention directs energy and information flow without our awareness. The direction of that information processing is still being shaped by the mind; it simply is not a part of our conscious subjective experience of knowing, of being aware.

The great news is that you can learn to sense these various aspects of attention, whether they are guided by you or pulled by things not directed by you; or whether they involve awareness and are focal, or do not and are non-focal. This directing of energy and information flow is *attention*. *Awareness* is the subjective experience of knowing within consciousness. We "know" what is going on around and within us, with the term "know" here meaning not factual knowledge but rather a subjectively felt texture of the present moment's unfolding. We can cultivate access to a more open experience of being aware, and this capacity for conscious choice and change empowers our lives to move with flexibility and intention toward a more integrated way of living. Training the mind is all about building these skills of attention, awareness, and intention.

Reflections:

Today I invite you to reflect on any experience you had today or recently of realizing that your awareness included content that was not directly inspired by the scene or setting in front of you. Perhaps a moment of sadness from your past coloring your view of the present moment? A feeling of irritation that seemed out of proportion to the actual situation in which you found yourself? Where did it come from? You may find it helpful in the weeks, months, and years ahead to make this kind of reflection a daily habit as you continue to bring the benefits of your Wheel practice into your daily life.

Our nonconscious mind has a profound impact on what we do with our behavior, and on how we feel and think, even when we are not aware of these influences of non-focal attention on our mental lives.

day 17—reflection on today's wheel practice

Today, as you moved through the relational segment of the Wheel, did any particular person or persons come to mind? Who were they and how would you describe the emotions that emerged as you turned your focal attention toward them and opened your awareness to them?

day 17—exercise: forgiveness

Today I encourage you to find the time to do an additional meditation session that expands on the "kind intention" portion of your Wheel practice earlier in the day. This time, the focus will be on offering forgiveness for whatever pain or harm someone else may have caused you in the past; and then asking yourself for forgiveness for something you may have done to cause pain or harm to someone else. Forgiveness is not stating that what happened was right or good; forgiveness, as my colleague and friend Jack Kornfield has suggested, is giving up all hope for a better past. As you take this time to feel forgiveness toward another and offer forgiveness to yourself, you can repeat the following phrases quietly in your mind or even say them softly aloud:

First, take a breath and say:

May all living beings be happy . . . May all living beings be as healthy as they can be . . . May all living beings be safe . . .

And may all living beings flourish and thrive.

Second, taking a deeper breath, say to the person you are extending forgiveness to:

May you be happy . . .

And live with meaning, connection, and equanimity . . . And a playful, grateful, and joyful heart.

May you be as healthy as you can be . . .

And have a body that gives you energy and flexibility . . . Strength and stability.

May you be safe . . .

And protected from all sorts of inner and outer harm.

And may you flourish and thrive . . .

And live with the ease of well-being.

Last, allow yourself to forgive yourself for something you have done or said that caused another person pain:

May I be happy . . .

And live with meaning, connection, and equanimity . . . And a playful, grateful, and joyful heart.

May I be as healthy as I can be . . .

And have a body that gives me energy and flexibility . . . Strength and stability.

May I be safe . . .

And protected from all sorts of inner and outer harm.

And may I flourish and thrive . . .

And live with the ease of well-being.

day 18—reflection on today's wheel practice

In today's Wheel practice, as you turned your attention to mental activities on the third rim segment, did you notice any "gaps" between one mental activity and the next? What did that space feel like? Can you describe it or even try to draw it below?

day 18—connecting to the collective mwe

Our self is not only what exists in the inner life of the body—the "I" of who we are. We are also part of an interconnected whole, part of a relational "we." But how can we integrate this differentiated bodily "me" or "I" with a relational "us" or "we"? Integration is the honoring of differences and then their compassionate, respectful linkage. If we integrate "me" and "we," we come up with an integrated identity, one that we can name with a new term: "MWe." Extending today's Wheel practice, here is an additional short meditation—you might try it for five or ten minutes—that will focus your awareness on the collective MWe.

Find a comfortable place to sit or lie and focus on your breath. Now let your visual focus of attention scan over your body—the source of your inner identity as an "I" or "me." Now let your visual attention widen to take in the world around you—other people who might be nearby, the room, or the natural world you might be in. Now let your mind take in the images of all the people you know, and now all human beings, to our whole human family. Expanding this lens of identity, letting your relational belonging broaden, see if you can sense your being a part of all of nature—you belong to all people and the planet. Your inner identity is me, your relational identity is we, and together your integrated identity is Me plus We equals MWe.

Now that you've used your mind's lens to move from a close-up focus to a wide-angle view on identity, you can take this inner and relational self inside and offer positive wishes of kind intention to MWe:

Repeat the following phrases aloud or say them silently in your mind as you continue to breathe deeply:

May MWe be happy . . .

And live with meaning, connection, and equanimity . . . And a playful, grateful, and joyful heart.

May MWe be as healthy as MWe can be . . .

And have bodies that give energy and flexibility . . . Strength and stability.

May MWe be safe . . .

And protected from all sorts of inner and outer harm.

And may MWe flourish and thrive . . . And live with the ease of well-being.

Reflections:

How did it feel to widen your identity lens and focus on the collective MWe of an integrated identity and broadened belonging? Did any particular faces of individuals you know or know of come into your mind? Did your heart move toward any communities in particular, either that you are a part of or know only from reading about or seeing images on television? What did MWe feel like to you?

day 19—reflection on today's wheel practice

day 19—exercise:
open awareness

Exercise:

Today I'd like to suggest that you do a deep dive into the hub it-self. As part of your Wheel practice today, after the third segment review, add the step of imagining bending the spoke of attention around 180 degrees, aiming attention right into the hub itself. Others find it more useful to imagine sending the spoke of atten-tion out a bit and then retracting it into the hub. Still others prefer to simply leave the spoke of attention inside the hub, or to have no spoke at all and rest in the knowing of the hub of the Wheel. Whatever way works best for you, the idea is the same: Open the hub of the Wheel to the experience of becoming aware of awareness, to simply drop into pure awareness, to simply be aware. Let's begin this practice right now for three to five minutes.

Reflections:

What was this hub-in-hub experience like for you? Many of my patients and workshop participants have found that it felt odd, at least at first. Did it feel disorienting or confusing for you? For some, there is simply a sense of becoming lost, of spacing out, of not knowing what to do. No need to worry. In many ways, this

step of resting awareness in awareness, of being aware of awareness, is quite advanced. I once taught this to someone who ran a
meditation center; he said that in his forty years of teaching, this
was, for him, the most advanced stage of his practice.

day 20—reflection on today's wheel practice

As you continue this journey with the Wheel, you may find that some days leave you with a new tool that you can then take into the following days. Perhaps a word that emerged in your mind that helped you return to the hub when you had become distracted? A sense of clarity about how to live from the hub's spaciousness? A way to be kinder to yourself and others, realizing we are all in this journey of life together? A metaphor for the mind that brought you closer to understanding how to become more and more present in your practice and your life beyond?

Recognizing these experiences may support you in findings even more ways each day to access the hub of open awareness and live life with more fullness and freedom.

---

day 20—a new tool:
the consolidated wheel practice

The Wheel of Awareness has a certain rhythm and fullness to it that many have wanted to preserve and yet find a way to do the Full Wheel practice within shorter session times. For this reason, I created the Consolidated Wheel, which is the Full Wheel, just done at a faster pace. Today I invite you to give the Consolidated Wheel a go and see if it might be a helpful tool for you in the days, weeks, and years ahead.

The basic idea is this: With each movement of the spoke of attention, we are harnessing focal attention skills to focus the spoke, take in the rim elements, and then move the spoke as we redirect attention. In the first segment, the first five senses— hearing, sight, smell, taste, and touch—are taken in with each in-breath and out-breath cycle. The key approach here is to move the spoke at the timing of your in-breath, soaking in the rim point with the in- and out-breaths, and then shifting focus on the next in-breath.

For the second segment of the rim, it can be an added feature to imagine the sensation of the breath coming in from the part of the body you are focusing on and then exhaling and sending the breath in the direction of the next part of the body you will focus on, and then taking in breath as you focus on this next part of the body. Of course, if there is a part of your body that needs a little extra breath, a little extra attention and time, go for it. Take whatever time you need, and whatever time you have.

In order to cultivate more open awareness, on the third segment of the rim I find a few cycles of the breath seem to work well for the two steps of that rim review exploring mental activities. And then for the bending or retracting of the spoke part, for the hub-in-hub step, I give myself whatever number of breaths I need. Sometimes for this hub-in-hub part of the practice, I lose track of the breath altogether, so it's helpful to set a timer if I am on a schedule so I'm not late for my next appointment. I may set a three-minute interval bell to ring before the ending tone so I can get the fourth segment review in without rushing through the end of this Consolidated Wheel practice.

On the fourth segment, it becomes very natural again to time the shift of focus from those closest to us, outward to all living beings with each breath cycle, in and out. And then when it comes to the statements of kind intention, you can try out many ways to fit each set of phrases into the cycle of the breath. One way I find very soothing is to make the initial statement with the in-breath and then the details on the out-breath. Then I repeat the basic phrase—for example, "May all living beings be happy"—at the end of the out-breath. Here it might be something like this:

(in-breath): May MWe be happy

(out-breath): and live with meaning, connection, and equanimity (in-breath): and a playful, grateful, and joyful heart.

(out-breath): May MWe be happy.

In this way, you can play with the timing of these statements to find something that suits both you and the particular phrasing.

day 21—reflection on today's wheel practice

You have come to the end of your journey through this 21-day deep dive into the Wheel of Awareness practice. Send your mind back to the first day's practice of the Basic Wheel. What insights have you gained into the architecture of your own mind since that first day that might help guide you as you continue to deepen your practice in the months to come?

day 21—exercise: write your own guided wheel practice

Over the past weeks you have likely committed some of my guided "instructions" for the Wheel practice to memory. You have also likely developed your own inner compass that guides you just as surely, if not more. In the space below, I invite you to write your own "instructions" using the entirely personal signposts that guide you as you move through your personal Wheel practice. This is your practice, your time with your mind, and making it your own in whatever ways that work is wonderful! Perhaps the rim "segments" for you are not segments at all; perhaps they are regions . . . or mental landscapes. Perhaps you see the image in three dimensions, more as a sphere, and this is your sphere of awareness. In your own inner language, describe your journey through the Wheel.

the wheel of life

I hope that over the course of the past three weeks, you have discovered a personal practice that will help you feel present, connected, and aware in each and every moment of your life. You've learned a way to bring more integration into your life, a practice for health across your lifespan. Remember:

Take your time,

Feel your life,

Breathe into the Wheel.

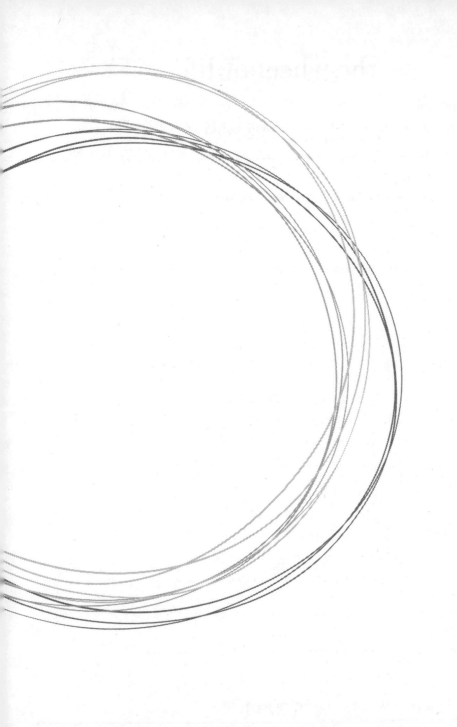

Resources

BOOKS

Brewer, Judson. *Unwinding Anxiety: New Science Shows How to Break the Cycles of Worry and Fear to Heal Your Mind.* New York: Avery, 2021.

Siegel, Daniel J. *Aware: The Science and Practice of Presence.* New York: TarcherPerigee, 2018.

——. *IntraConnected: MWe (Me + We) as the Integration of Belonging and Identity.* New York: Norton, 2021.

——. *Mind: A Journey to the Heart of Being Human.* New York: Norton, 2017.

Welch, C. S. *The Gift of Presence: A Mindfulness Guide for Women.* New York: TarcherPerigee, 2020.

WEBSITES

The Center for Compassion and Altruism Research and Education, Stanford University—http://ccare.stanford.edu

Center for Healthy Minds, University of Wisconsin-Madison—https://centerhealthyminds.org

Dr. Amishi Jha's Lab—http://www.amishi.com/lab

Dr. Dan Siegel's Resources—http://www.drdansiegel.com/press

Dr. Elissa Epel's Lab—http://www.amecenter.ucsf.edu

Dr. Judson Brewer's Lab—https://www.brown.edu/academics/public-health/research/mindfulness

Greater Good Science Center, University of California, Berkeley—https://greatergood.berkeley.edu

Mind & Life Institute—https://mindandlife.org

Mindfulness-based cognitive therapy—http://mbct.com/

Mindsight Institute online courses and workshops—https://www.mindsightinstitute.com

National Center for Complementary and Integrative Health—https://nccih.nih.gov

UCLA Mindful Awareness Research Center—http://marc.ucla.edu

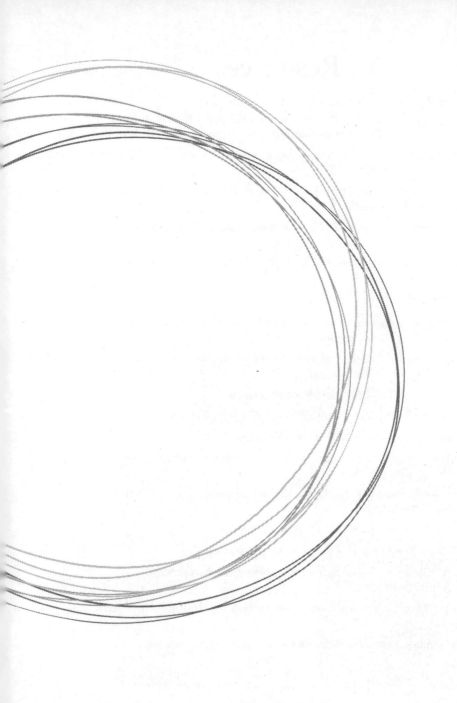

also by

Daniel J. Siegel

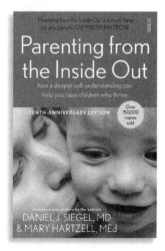

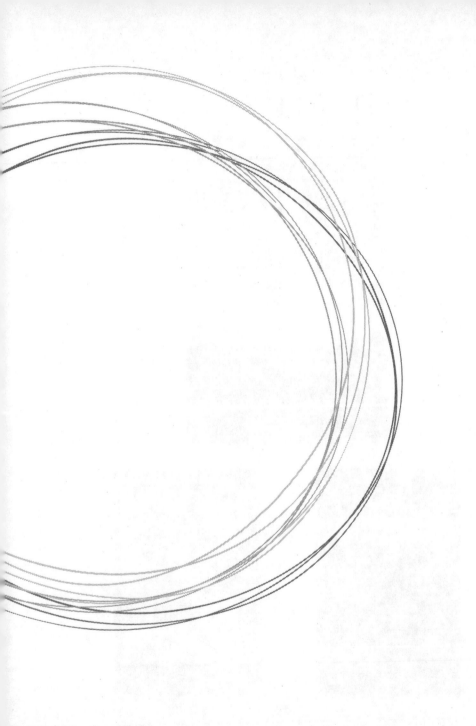